eye to eye™ BOOKS

NIGHT CREATURES

photographs by
Simon M. Bell

text by
Andrea Holden-Boone

SOMERVILLE HOUSE, USA
NEW YORK

ISBN: 1-58184-006-3   A B C D E F G H I J

Illustrated by Julian Mulock
Art Director: Neil Stuart
Design: FiWired.com
Printed in Hong Kong

Somerville House, USA, is distributed by
Penguin Putnam Books for Young Readers,
345 Hudson Street, NY, 10014

Published in Canada by
Somerville House Publishing
a division of Somerville House Books Limited
3080 Yonge Street, Suite 5000
Toronto, Ontario M4N 3N1

All photographs by Simon Bell with the exception of the
following: Earth from space, page 4: NASA/BPS; Earth
from space, page 5: NASA/BPS and Jon Lomberg; luna
moth, page 13: Doug Wechsler.

The Publisher and photographer would like to thank the
following organizations and individuals without whose
help this book would not have been possible: Toronto
Zoo, Wildlife on Easy Street, Northwood Buffalo and
Exotic Animal Ranch; special thanks to Pat Quillan,
Norm Phillips, Wayne Jackson, and John Carnio.

CONTENTS

Words that appear in **bold** are explained in the glossary.

AS DIFFERENT AS NIGHT AND DAY

3-D Jaguar

3-D Sugar Glider

For many humans and animals, night is the time to rest and sleep. But for some animals, the beginning of night is the signal to wake up and get moving. Animals that sleep during the day and wake up at night are called nocturnal animals. They are the creatures of the night.

Earth's Endless Spinning

The sun seems to disappear at night and then return the next morning. Why does this happen? The Earth is like a giant, spinning ball. It takes 24 hours for the Earth to complete one rotation. It is this turning that creates night and day. The side of the Earth that faces the sun has daylight, while the other side is in darkness. When our side of the Earth turns away from its glowing sun, daytime comes to an end and night begins. So the sun doesn't ever disappear — it's just shining on another part of the world.

Nighttime happens at different times depending on the continent. While North Americans are getting ready to go to sleep, people in Australia, for instance, are about to have their morning alarm clocks wake them up.

Humans and animals don't feel the spin, however, because everything — the land, water, and atmosphere — is moving together. What humans and animals do feel and see is the difference between night and day.

▲ As Earth rotates away from its sun, daytime comes to an end and night begins.

► The moonlight that helps night creatures find their prey is actually reflected light from the sun.

The Night Shift

The Earth's **resources** are in high demand. Because of this, some animals have adapted to take advantage of the night. These nocturnal animals don't have to compete with day-active animals for food. Some creatures become nocturnal because their prey is nocturnal. They develop special adaptations to make them successful hunters.

Some animals that live in hot, tropical regions of the Earth prefer to hunt at night, when it's cooler. As the night air cools, moisture condenses, causing **dew** to form. The hot sun is not around to dry up the **atmosphere**. Creatures that must keep their bodies moist emerge at night.

Diurnal — day-active — animals seek out their dens, nests, or **lairs** to curl up in and sleep in during the night.

▼ **Earth is the third planet from the sun, situated approximately 92.96 million miles (149 million km) away.**

Why can stars be seen only at night? Actually, Earth has a star that shines during the day… the sun! The sun is the star that's closest to the Earth, and so more of its light reaches us. Distant suns, or stars, may be as bright as ours, but, because they're so far away, little of their light reaches us. When the sun is in the sky, its light is so powerful that it drowns out the light from the distant suns.

NIGHT SONGS

▼ The spring peeper is a small, brownish North American frog whose incessant call fills the night air.

The night has a music all its own. Animals use sound to defend their territories, find mates, and keep track of their young. Whether it's lions roaring to each other at dusk on the African plains or a crowd of birds calling to each other during migration, listening to their sounds is one way to learn about the creatures of the night.

Buzz and Chirp

Insects create a symphony of noise at night. One common sound is the chirp of the cricket. The males make this noise by rubbing their wings together. They hope a female cricket will appreciate the serenade and seek out the male to mate. Some scientists claim that crickets tell the temperature. The faster they chirp, the warmer the night.

Another sound commonly heard at dusk, and also during the day, is the song of the katydid. The katydid has a low, buzzing call that sounds just like its name: "Katy did, did, didn't."

Croak and Peep

Toads and other species of frog (see White's-tree-frog card) add their own tune to the nighttime symphony. Bullfrogs croak by forcing air from their lungs into a sac underneath their chin. The sac blows up and then the air is released in a deep, resounding call. In wet, marshy areas, a sound like millions of shrill peeps can be heard. The noise comes from tiny tree frogs called spring peepers.

Bird Songs

After diurnal birds **roost** for the evening, nocturnal birds can be heard. Even though it is not completely nocturnal, the nightingale — known for its beautiful song — is best heard at night. Owls hoot, screech, or wail, depending on the species. Nighthawks squawk, marsh bitterns boom, and other birds call if they are disturbed by a **predator**.

Finding Each Other

Cats often howl at night to find mates or let other cats know when they are trespassing on claimed territory.

Singing to the Moon

Coyotes and wolves (see wolf card) are famous for howling at night. Their chilling yet beautiful calls gather the pack together and define their territory to **intruders**. These animals seem to enjoy their vocal exercise — yipping, and howling to each other.

▲ For many people, the call of a wolf defines the northern wilderness.

▲ Lions roar at night to celebrate a kill, defend their territory, or claim a mate.

Grasshopper Mice

Grasshopper mice that inhabit the deserts of the United States are fierce little creatures that hunt by night. Their howls are really **ultrasonic** squeaks that people can't hear. Other grasshopper mice can hear the howls, however, which are used to attract mates and define their territory.

MOVING AND CHANGING

N ight is a time of movement and change for many creatures. Even some diurnal animals take advantage of the darkness and the protection from predators that it provides.

Celestial Navigation

Early explorers traveled to new lands with the stars as guides. Recently, scientists have discovered that birds use a similar method to stay on track at night. Birds can actually learn to read the stars and use this nighttime map to migrate.

An Internal Compass, Too!

Some birds are born with magnetic crystals inside their skulls. These crystals act like a compass. By following the stars and using their magnetic sense of direction, birds are able to make their round-trip journeys.

A Night Mission

Though turtles don't migrate with the changing seasons, baby tropical deep-sea turtles have to find their way home — for the first time. Each year, the female turtles swim to shore to lay their eggs, then head back to the ocean. Several weeks later, the baby sea turtles hatch at night. The turtle hatchlings must make their way to the safety of the water as quickly as

▲ Once a year, female green sea turtles must journey to where they were born, go ashore to dig a nest in the sand, lay their eggs, and cover them up before returning exhausted to the sea.

▶ Instinct makes these turtle hatchlings scramble for the safety of the sea.

◀ Like a frog, an American toad has very porous skin and it can get the water it needs to survive by simply soaking its body in a pond or stream.

possible. By instinct, they follow the moonlight and starlight reflected in the ocean. Sadly, few baby turtles succesfully complete the trip. Most are eaten by predators, such as crabs and birds, before they ever reach the water.

Sometimes the babies mistake a human house for the ocean. Young turtles instinctively look for the light on the ocean, but too many artificial lights distract them. Now, people who live near turtle-breeding areas are asked to keep their lights low or off at night during breeding season.

Changing

During the night, moths come out of their **cocoons** and butterflies emerge from their chrysalises (protective coverings). Their bodies are soft and moist, and the damp night air prevents them from drying out. Even **crustaceans,** like shrimp and lobster, shed their shells underwater at night.

Toads (see marine-toad card) mate only where they were born. This proves dangerous for many toads in Great Britain and people have stepped in to help. On dark, rainy nights, thousands of toads cross highways and roads to travel back to their breeding ponds. Many are squashed under the wheels of cars and trucks. A "Save the Toad" campaign has begun where "toad fanciers" venture out at night as volunteer crossing guards for the migrating toads. These dedicated people stop traffic, scoop toads into buckets, and carry them across the highways. Injured toads are rushed to wildlife rescue centers.

NIGHT DETECTIVES

3-D Eyelash Viper
3-D Shovelnose Catfish

Animals that hunt by day rely heavily on their vision to track down prey. At night, it's a different story. Nocturnal creatures use a variety of senses to survive.

Vibrations Are a Dead Giveaway

Some animals are highly sensitive to **vibrations**. If danger lurks too close, gerbils will stamp their feet to warn others of predators. A spider knows if an insect is caught in its web by the jiggling of the silk strands. The trap-door spider can feel the tiny vibrations of a bug moving near its **burrow**. A cockroach's sensitive legs pick up vibrations that alert it to the movement of enemies — like humans — before they are seen.

Heat Sensors

The rattlesnake (a type of pit viper) has a tiny hole, lined with heat-sensitive cells, below each eye. These are the viper's "pits" and they enable the snake to locate warm-bodied animals.

Smelling and Tasting

Some insects, like the moth, use their **antennae** for smelling and tasting. Snakes test the air for scents by breathing in through their nostrils and by flicking out their tongues. Snakes can actually taste nearby prey.

The albatross and the shearwater use smell to detect fish swimming underwater. The oilbird of South

◀ **Cockroaches, like these Madagascar hissing cockroaches, have large, highly sensitive compound eyes, and a light suddenly turned on will send them scampering away.**

America sleeps by day and feeds on oily palm fruits by night, smelling the fruit in the dark.

Touching

The whiskers of a cat and those of a catfish function in almost the same way, even though one set of whiskers is in the air and the other is underwater. A cat's whiskers are attached to touch-sensitive nerve endings and help it to feel its way through the dark in search of prey.

Catfish whiskers (called barbels) do double duty. They can both feel and smell. The catfish (see shovelnose-catfish card) drags its whiskers along the river bottom in search of food and mates.

▲ The mysterious clouded leopard inhabits thick, forested regions of Asia. This small, wild cat lives and hunts in the trees. At night, the clouded leopard silently stalks prey through the branches of the forest — its keen sense of smell and eyesight help it to locate food in the dark.

Most animals, including humans, have two kinds of cells in their eyes: **rods** and cones. Rods are sensitive to light while cones are sensitive to color. Nocturnal animals have many more rod cells than cone cells — this allows them to see better in low-light conditions.

Night-active snakes and small wild cats have similar eyes. Both are almost color-blind and have pupils that close into vertical slits during the day. At night, the pupils can open very wide — almost the entire width of the eye — to allow maximum light through to the light-sensitive rod cells.

A NIGHT GARDEN

 Atlas Moth

▼ Usually, butterflies are more colorful than moths, but this uranid moth is exceptionally vivid.

Night Flowers

Most flowers bloom during the day and close up their petals at night. However, there are some flowers, like the nicotiana and moonflower, that bloom during the night. They take advantage of night **pollinators**, like the moth, lacewing, and firefly. Night-blooming flowers tend to be a pale color or white and seem to glow at night. Planting a garden while keeping the night bloomers in mind can create a magical space that will attract a variety of night-active insects and other animals.

The Moth

Moths (see atlas-moth card) are like butterflies, but come out only at night. Like butterflies, moths have wings that are covered with fine, feathery scales. Both the moth and the butterfly have a long, straw-like **proboscis** that is used for sucking up nectar from flowering plants. The moth's antennae are used for smelling and tasting. The antennae are so sensitive that a male moth can smell a female from one mile away.

Moths lay eggs that develop into caterpillars. When the time comes, the moth caterpillar spins a cocoon. Eventually a new moth emerges, and the cycle continues.

BELL JUNIOR HIGH SCHOOL

The Lacewing

Lacewings lay their eggs on the leaves and stems of plants. When the eggs hatch, the lacewing larvae come out to feed. Also called aphid lions, these larvae prey upon other small insects. The wily larvae camouflage themselves with leaves and twigs. They use this disguise to sneak up on unsuspecting prey.

▲ **The transparent wings of a lacewing are crisscrossed by veins, which make the wings look like they're covered in lace.**

Night Lights

The firefly (really a beetle) blinks in observable patterns on summer nights. Their eerie light is produced by a process called chemiluminescence — fireflies break down sugar to produce energy in the form of light. The males fly while the females stay on grasses or bushes. The female fireflies repeat the blinking patterns of the males to attract their attention. Then they mate.

▶ **The range of the luna moth extends from mid-western Canada toward the Atlantic coast and down to Florida and Texas.**

There is a spectacular species of North American moth called the luna moth. *Luna* is the Latin word for moon. This moth gets its name from the moon-shaped markings that decorate its beautiful pale-green wings. Like other moths, they fly at night and rest during the day.

UNDER COVER

3-D Marine Toad
3-D Bell's Horned Frog

Find a Little Bit of Night — During the Day

Flip over a large rock during the day and look at the dark earth underneath. A great deal of scampering will take place as bugs and other creatures make for cover. Salamanders, slugs, and snails that live under rocks and branches come out at night. These creatures need to keep their bodies cool and moist.

▲ This spotted salamander looks like a small, peculiar lizard. It is not a reptile, however, it is an amphibian. Like their cousins, frogs and toads, salamanders are able to live in water and on land.

The Early Bird

The expression "the early bird gets the worm" is true. Earthworms have soft bodies that easily dry out in the hot sun. During daytime hours, worms stay hidden under the earth — burrowing tunnels through the dirt. At night, when temperatures are cooler and the air is moist, worms often come to the surface. They venture out of their burrows to forage for food, especially when it's raining. Most birds sleep at night, so worms have a better chance of surviving out in the open when it's dark. But birds wake up early: their favorite time to feed is early dawn. The worms who have not retreated underground by this time are quickly snapped up as breakfast.

Snails and Slugs

Even though snails and slugs may look like insects, they're actually a type of **mollusk** called a gastropod. This funny name comes from the way they move. A gastropod, which means a "belly-foot," might look like it's sliding around on its stomach, but really it's inching along with the help of a big, muscular foot.

Like earthworms, snails need to keep their bodies moist. During the day they hide in their shells which protect them from the heat of the sun. They come out at night, when it's damp and dark, to eat.

Slugs are kind of like snails without shells. They live their lives coated in a layer of slime. The slime helps the slug to move, allowing it to slide along surfaces with its strong foot.

▲ A snail makes slow but steady progress using its one foot.

Nature's Pest Patrol

Toads (see marine-toad card) take advantage of the night to hunt insects. Though some are active during the day, most toads prefer the night. Like the earthworm, the toad also has porous skin and loses moisture easily. A dark, humid night is perfect weather for a toad. The common toad is a gardener's friend — eating as much as three times its weight in harmful insects every day. They hunt and eat just about anything, except the gardener's plants. Beetles, slugs, moths, centipedes, crickets, and cucumber beetles are just a few of the creatures that make up the toad's diet.

▲ The banana slug lives in the moist soils of northwestern American forests and it particularly likes redwood forests. It can grow up to 8 inches (20 cm) long and weigh up to a quarter of a pound (0.1 kg).

BEAUTY AND THE BEAST

3-D Emperor Scorpion
3-D Pink-toed Tarantula

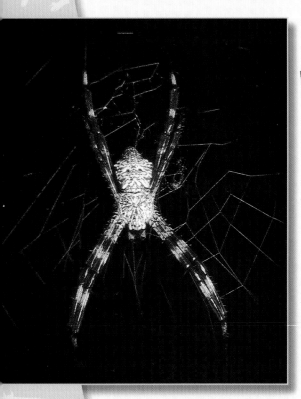

▲ If a spider, like this orb-weaver spider, senses movement on its web, it will quickly emerge, inject the trapped victim with poison, then carry its prey to safety. There, the spider will either eat its victim immediately, or wrap its catch in silk to store it for a later meal.

Night Weavers

Safe from most predatory birds, spiders venture out at night to spin their silky nets. All spiders have silk-producing glands called spinnerets on their abdomens. When spinning a web, a spider first climbs to an anchor: a tall blade of grass or a twig. Then it **secretes** a long strand of sticky silk and waits for a breeze to carry the thread to another fixed object where the thread attaches itself. This first strand is used as a base for the rest of the web.

▲ The most beautiful webs come from orb-weaver spiders. Their webs are intricate spirals that are well known for their design and strength.

Night Hunters

One of the largest nocturnal spiders in the world is the tarantula (see pink-toed-tarantula card). Its body is covered with fine, bristly hairs. Touching the hairs of this gruesome spider can sometimes cause a rash. The tarantula is a voracious meat eater, emerging at night to hunt insects, small frogs, lizards, and even other tarantulas. Some tarantulas are able to capture small snakes and birds and kill them with a venomous bite. A tarantula bite can also be poisonous to humans, but is not usually **fatal**.

Night Terrors — The Scorpion

The scorpion is a member of the spider, mite, and tick class. Some scorpions grow as large as six inches (15 cm), with the smallest being about one inch (2.5 cm). Larger scorpions tend to inhabit warmer regions of the Earth.

Like the spider, the scorpion has four pairs of legs and a pair of pincerlike appendages that look like the claws of a crab. Its slender, jointed tail has a stinging organ at the tip. While fighting or hunting for prey at night, the scorpion whips its tail over its back to inflict a sting. Then a gland containing **venom** contracts and pumps poison into the victim.

Some species of scorpion live up to 25 years. A few scorpions have venom that is 100,000 times greater than the poison cyanide, so their sting can be fatal to humans. In warmer regions, where scorpions are abundant, scientists go on nightly "scorpion hunts." Some species of scorpions are almost invisible by day. But scientists have discovered that the scorpions' bodies glow when the scientists shine a special "black light" on them. This way, scientists have been able to discover new species of scorpion, research their habits, and help people rid their homes of these night terrors.

◀ There are over 1,500 different species of scorpions, including this emperor scorpion (see emperor-scorpion card), which can be found in all parts of the United States and Canada, as well as other parts of the world.

Creeping Crustaceans

The Earth's oceans and seas are full of nighttime activity. Shrimp, crab, and lobster are more active at night. They **scavenge** the ocean floor in search of leftovers dropped down by higher-level feeders.

Sea Stars and Urchins

Sea stars and urchins hunt mollusks and scrape rocks and coral for **algae**. Some climb to the tops of coral reefs under the safety of darkness, seeking out food before daylight gives them away.

Microscopic Night Swimmers

Zooplankton, or tiny marine creatures, rise to the ocean's surface at night. Zooplankton are an important source of food for animals that filter feed — that is, strain ocean water through their bodies and sift out **nutrients**.

Coral animals and sea anemones are just a few of the feeders that depend on zooplankton. At night, coral "flower," or pop out their **tentacles** to trap microscopic animals floating by.

Large Predators

Sharks, eels, and barracuda are more active at night, especially at dusk when the slanting rays of the sun cast shadows on prey. These hunters try to catch diurnal animals as they hurry to their nighttime hiding places.

▲ Sea stars, sometimes called starfish, have five arms. If a sea star loses one of its arms to a predator, it will grow a new one.

Some animals don't glow but squirt streams of light into the ocean water. The deep-sea octopus, when threatened, escapes predators by ejecting such a glowing stream into the water behind it. This confuses predators and gives the octopus a chance to get away. Some crustaceans off the coast of Japan are **luminescent**. They have separate glands containing bacteria which, when squirted into the water, generate light.

Underwater Night-Lights

Some types of jelly fish produce green flashes. Certain species of anemone glow. The flashlight fish has glowing sacs of bacteria under each eye that it can turn on and off at will. These animals don't create light from stored energy from the sun, but rather, from chemical reactions. This is called bioluminescence — when chemical energy is transformed into light energy.

The Hawaiian Squid

This swimming night-light has a special cavity on its underside that collects bioluminescent bacteria from the sea water. At night, the cavity, or "light organ," on the squid's belly glows. This is thought to protect the squid from predators by casting a shadow in the opposite direction.

◀ **A flashlight fish glows in the depths of the ocean.**

▼ **Predator fish, like this black-tip shark, use the setting sun to help them catch prey. As the sun sets, it creates long shadows of fish and other creatures underwater. The shadows of potential prey are more easily spotted by the shark at dusk, rather than when the sun is directly overhead.**

LIVING IN THE DARK

3-D **Naked Mole-rat**
3-D **Northern Goshawk**

Naked Mole-rats — 24-Hour Night Creatures

Naked mole-rats spend their entire lives underground in the hot, dry regions of Africa. These rodents look a little like unfinished mice. They are about the same size as mice, but are almost completely hairless, with wrinkled pink skin.

The naked mole-rats (see naked mole-rat card) rely on smell and sound, instead of sight. They keep track of each other by smell alone.

Home Sweet Home

The mole-rats dig burrows and tunnels that can stretch up to two miles (3 km) long. Naked mole-rats are adept at scurrying both backward and forward through the tunnels. The few hairs they do have are highly touch sensitive — like whiskers — and help the mole-rats feel their way around.

Working Together

Naked mole-rats live in colonies that have a complicated social network — much like bees — with workers, soldiers, and queens.

Each mole-rat has a job to do. Some are in charge of extending the tunnels. These diggers work as a team, passing the excavated earth up a line to dump it outside the tunnel entrance. Some naked mole-rats are cleaners. They travel the corridors picking up and disposing of debris. Other mole-rats are assigned to search for food, collecting **tubers** that grow underground. The larger males defend the colony from invaders.

◄ **Naked mole-rats live in colonies that can have up to 300 individuals. If the usual 85° F (29°C) temperature of their burrows and tunnels drops, they will huddle together for warmth**

▲ Although mole-rats may look strange to us, the important thing is that they are attractive to other mole-rats.

Naked mole-rats' teeth are so strong they have been known to dig through concrete. Zookeepers have found that plastic tubes are able to contain the mole-rats. Because naked mole-rats are so sensitive to sound, zookeepers have to insulate the colony's network of tunnels in order to keep the noise of human footsteps and voices to a minimum. Otherwise, the mole-rats become confused and act aggressively.

Queen of the Rats

Each mole-rat society has a queen. She is larger than the rest of the rats and is responsible for bearing the young. When the queen is pregnant and large, worker mole-rats actually help push her through the tunnels. When the queen dies, another female takes her place. The new queen grows larger and her spinal column lengthens. Scientists still are not sure what triggers the change.

Other Burrowing Relatives

Moles and voles also live underground. Like the naked mole-rat, they dig burrows and tunnels and eat roots and tubers. The starnose mole has tentacles around its nose that move continuously in search of food.

NIGHT PROWLERS

3-D **Leopard**

3-D **Margay**

3-D **Raccoon**

3-D **Brush-tailed Bettong**

Secretive Cats

Most small cats are nocturnal. Even the common house cat is awake at night, much to the irritation of its owners.

Cats are well-adapted to the night. Their powerful sight, smell, and hearing help them hunt prey. Cats can see six times better than humans at night.

Some small wild cats, like the margay (see margay card), are arboreal: they spend most of their time in trees. Chasing prey through dark forest branches requires quick reactions and a good sense of balance.

Sly Foxes

Foxes are related to dogs, and are known for their cunning behavior. The fennec fox lives in the Sahara desert. It spends the daylight hours, when the desert is the hottest, in its underground burrow. At night, the fennec fox comes out to hunt. Sharp hearing helps the fennec fox locate scurrying prey, such as scorpions, desert lizards, and snakes in the dark.

The New World red fox inhabits woodland areas. The kit fox, its relative, lives in semidesert regions. Both are sly, nocturnal hunters. Like the fennec, these foxes use their hearing to seek out food, such as crickets, grasshoppers, rodents, and rabbits in the dark.

▼ **Known for their cunning behavior, some red foxes know how to break their scent trail by entering running water, such as a stream, then swimming or slogging upstream.**

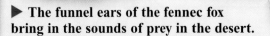

▶ **The funnel ears of the fennec fox bring in the sounds of prey in the desert.**

▶ **If your family pet has been foolish enough to tangle with a striped skunk, the best remedy is an old one — bathe the dog in tomato juice!**

The Raccoon

Raccoons (see raccoon card) are often thought of as "pests" that plunder gardens and tip over garbage cans in search of food at night. These "masked bandits," however, are cunning night creatures known for their **dexterous** forepaws. In fact, the word raccoon comes from a Native American word meaning "he who scratches his face with his hands." Raccoons can pry just about anything open and like to moisten their food before they eat it. The raccoon prefers to live in wooded areas, but many are found thriving in cities where they make a living from garbage waste, small animals, and plants.

Smelly Skunks

Skunks are related to weasels. They come out at night to forage for food, such as fruit, plants, rodents, insects, and reptiles. Unfortunately, skunks are fond of scavenging through people's garbage.

Many dogs have learned a hard lesson from quarreling with a skunk. The skunk has a unique defense mechanism. The skunk has two special glands full of a yellow, oily liquid that has a strong, nauseating odor. If threatened, a skunk ejects this liquid. The skunk can hit a target 12 feet (4 m) away. If this vile liquid comes in contact with the eyes of an enemy, it can cause temporary blindness and a lot of stinging.

The only predator that doesn't seem to mind the smell of the skunk and is not affected by its spray is the great-horned owl.

23

▶ **The southern hairy-nosed wombat's teeth have no roots and grow continuously. They need to replace dull teeth worn out from grinding and chewing food.**

Close Cousins — The Wombat and Koala

The wombat and koala are related **marsupials** that inhabit different regions of Australia.

The Hairy-Nosed Wombat

The hairy-nosed wombat (see southern hairy-nosed wombat card) spends the day in its burrow. At night, the wombat emerges to sniff out its dinner of leaves, grasses, and herbs.

Wombats can weigh up to 40 pounds (18 kg). They have short, stocky legs and a slow shuffling way of walking. Their bodies are rounded and covered with a short, fine coat of fur that ranges from dark silver to brown, and they have short tails.

The Koala

The koala, sometimes called the koala bear or native bear, lives mostly in trees and is an excellent climber.

The koala sleeps most of the day in a forked branch, then feeds on the leaves of the eucalyptus at night. Eucalyptus leaves contain an oil that makes them poisonous to most other animals. The koala's liver, however, can eliminate the poison from eucalyptus leaves.

Baby koalas are born one at a time and spend over seven months in their mother's pouch.

The Tasmanian Devil

As night falls, the Tasmanian devil (see Tasmanian-devil card) ventures from its burrow to hunt food. The Tasmanian devil is the largest surviving **carnivorous** marsupial and can weigh up to 18 pounds (8 kg). It is covered with short, black fur and has a white band of color around its neck.

The Tasmanian devil's claim to fame is its powerful jaws. It has been known to kill and consume animals many times its own size. The devil eats every part of its kill, including the hard skull bones.

The Spiny Anteater

The land-dwelling echidna (see short-beaked echidna card), also called the spiny anteater, is a member of the monotreme family (the only mammals known to lay eggs). Echidnas sleep in burrows by day and forage for food at night. Long snouts and long, sticky tongues enable them to root for ants and termites, then lap them up.

▲ The koala is not a bear at all. It is a marsupial like a possum or kangaroo.

▶ The female echidna lays one egg at a time and deposits it in her pouch. After ten days, the youngster hatches.

25

OWLS

3-D Barn Owl
3-D Eagle Owl

Owls are the silent hunters of the night, preying on all kinds of small animals and insects. Owls inhabit nearly all regions of the Earth and range in size — the smallest being the elf and pygmy owls at only 6 inches (15 cm) tall. The largest owls, including the great-horned owl of North America, can reach heights of two feet (60 cm) with wingspans of five feet (1.5 m).

Wings and Feathers

An owl's body is designed for silent flight. While most birds have feathers with sharp edges, the owl's feathers have a fluffy fringe that muffles sound. Its long wingspan gives the owl agility and speed. It can dive, swoop, circle, and fly in a straight line with ease.

Talons

The owl's deadliest weapons are its razor-sharp, hooked talons. Each of the owl's four toes is equipped with a clawlike talon. These talons help the owl cling to its perch and grab prey securely. The owl can kill its prey instantly with the pierce of a talon.

Night Vision

An owl's eyes are set in a forward position and cannot move. Because of this, the owl's neck muscles are very flexible, allowing it to turn its head in a complete circle, or even upside down. Though owls are color-blind, they are able to distinguish shapes in very low light.

◀ **The extended feathers on top of its head give this owl its name — the long-eared owl. It raises and lowers its feathers depending on its mood. The feather tufts are not true ears. The long-eared owl's real ears are hidden by the feathers that surround its face.**

◀ The fluffy edges on this owl's wings allow it to fly silently and to surprise its prey.

▼ Unlike other owls, this barn owl (see barn-owl card) has a distinctive heart-shaped face.

Acute Hearing

Even when an owl is perched high in a tree, it can detect scurrying rodents on the ground. One of the owl's ears is a bit lower than the other, which helps the owl determine the distance and direction of the noise it hears.

On the Hunt

When scanning for prey, the owl selects a high perch where it can see for a distance. Once prey is heard or spotted, the owl silently flies in for the kill, swooping down and grabbing its victim with its talons.

Owls usually swallow their prey whole. After a meal, an owl will **regurgitate** the indigestible parts of its dinner: bones, feathers, and fur. Scientists study these regurgitations to learn about an owl's diet.

Don't get too close to an owl's nest. Owls are known for viciously defending their territories, especially when chicks are present. If threatened, owls show a variety of behaviors just before they attack. Some puff up and stretch their bodies, making them look twice as big as they actually are. Others hiss and screech. The burrowing owl makes a sound like that of a rattlesnake. All these actions are strong warnings to stay away.

BATS

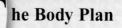

3D Gray-headed
Flying Fox

3D Vampire Bat

▼ **Using echolocation, a bat finds dinner.**

The Body Plan

The bat's hind feet are equipped with five claws that allow it to hang from branches and rocks. Its wings are really an extension of its front legs, which are covered with a flexible skin membrane. Bats (see gray-headed flying fox card) flap their wings in a rolling forward-then-backward motion and are capable of reaching great speeds.

Hanging Around

Some species of bats roost in caves, where they spend the day hanging upside down and resting. Bat colonies can have as many as 30 million individual bats. At dusk, bats stream out of their roosts in swarms, flooding the night sky. Other species, like the gray-headed flying fox, roost in trees, where they congregate in large numbers.

Time to Eat

Some bats eat only fruit and nectar, while others are insect hunters. Fruit bats have long snouts and tongues for feeding on plants. Scout bats sniff out ripening fruit, then lead the rest of their colony to feed.

Echolocation

All bats have eyes, but they don't use them much. Insect-eating bats have a unique system for locating prey. These bats send out rapid, ultrasonic signals that bounce off anything in their path. These sound signals travel back to the bat's sensitive ears, allowing the bat to tell the direction and speed of nearby prey. In this way, the bat is able to pinpoint its victim's location.

Some moths have developed the ability to jam bat sound waves. These moths squeak back at the bats. This misdirects the bat's sonar during an attack.

The Real Vampires

Vampire bats, found in Central and South America, have a somewhat ghoulish reputation. They are small — only two to four inches (5 to 10 cm) long — but have needle-sharp teeth. Horror movies have shown these bats attacking humans and devouring their blood. In reality, however, vampire bats usually attack farm animals who sleep out in the open at night. The vampire bat's saliva contains a substance that keeps its victim's blood flowing while the bat is feeding.

The sugar glider (see sugar-glider card) is a marsupial that looks like the flying squirrel. The sugar glider has flaps of loose skin on each side of its body. It glides from spot to spot by jumping with its hind legs, then spreading its four legs out into "wings." Though the nocturnal sugar glider can "fly" up to 165 feet (50 m) in this way, it is not capable of sustained flight.

▲ **The Australian gray-headed flying fox is one of the largest bats in the world. It can weigh over two pounds (1 kg) and have a wingspan of 3.5 (1 m).**

◀ **Flaps of skin help carry the sugar glider through the air, somewhat like a kite.**

STEREO PHOTOGRAPHY

Our Eyes

A pair of eyes is one of the most complex systems found in nature. Scientists don't fully understand all the intricate mechanisms that allow humans and animals to see. Scientists do know, however, that both eyes work together with the brain to form images that have three dimensions: length, width, and depth. Viewed through only one eye, everything appears flat, or two-dimensional. Using two eyes allows humans to perceive the third dimension — depth.

How the Eyes Work

Because our eyes are a few inches apart, each eye sees a slightly different angle of the same object. The information from each eye is carried by nerves to the brain. Then, in a process called fusion, the brain forms a blended image that is three-dimensional. Fusion allows humans to judge distances between objects, and determine how far away they are.

How a Stereo Camera Works

Stereophotography works like human eyesight. The most sophisticated stereo cameras have two lenses about the same distance apart as human eyes. Two images are taken simultaneously, each with a slightly different angle of the same object. When both images are viewed through a special viewer, called a stereoscope, the two images are blended together to become one image that has three dimensions.

Eye-to-Eye™ Books

This Eye-to-Eye™ book contains cards with paired images of night creatures taken by a stereo camera. When you look at the cards through the stereoscopic viewer, a 3-D image is formed in your brain. This is because the left side of the card mimics what the left eye might see, while the right side mimics the right eye's perspective.

To View the Cards

Carefully remove the viewer from the front of the book. Lift flap and insert tab. Carefully remove the cards from the back of the book. Insert cards one by one into the slot. When you have finished viewing the cards, store the viewer and cards in the pocket on the inside back cover.

Simon M. Bell

specializes in stereographic nature photography and is founder and president of BPS, a multimedia studio based in Toronto, Canada.

Simon began shooting pictures at the age of six when his father gave him a "Brownie" box camera.

GLOSSARY

Algae - plants that do not have true roots, stems, and leaves but often have green coloring.

Antennae - the long, thin feelers on the head of some animals.

Atmosphere - the climate of a place.

Burrow - a hole, tunnel, or opening dug in the ground by a small animal such as a mole or rabbit.

Carnivorous - feeding on the flesh of other animals.

Celestial - of, or having to do with the sky or the heavens.

Cocoon - a covering of silky strands made by a caterpillar.

Cone - one of the short sensory end organs of the vertebrate that function in color vision.

Crustaceans - any of a group of animals with a body that has a hard outer covering.

Dew - small drops of water that form from the air and collect on the surface, usually during the night.

Dexterous - having or showing skill in use of the hands.

Fatal - causing death.

Incessant - never stopping.

Instinct - an inner feeling or way of behaving that is present at birth and is not learned.

Intruder - one who comes unasked and unwanted.

Lair - the den or home of a wild animal.

Luminescent - capable of producing light by a process other than incandescence.

Marsupial - one of a group of animals that live mostly in Australia. The females have a pouch on the outside of the body. The newborn young are carried in this pouch.

Mollusk - one of a large group of animals that have a soft body and usually live in water. Most mollusks have a hard outer shell. Snails, clams, and oysters are mollusks with shells. Octopuses, squids, and slugs are mollusks with no outer shell.

Nutrient - something that nourishes.

Pollinators - those that carry pollen from the stamens of one plant to the pistols of another.

Porous - full of tiny holes permeable by water, air, etc.

Predator - an animal that lives by hunting other animals for food.

Prey - an animal caught or hunted by another animal for food.

Proboscis - any long, flexible snout adapted for piercing or sucking.

Regurgitate - bring undigested food back from the stomach to the mouth.

Resources - those things that can be used for support or help.

Rods - one of the microscopic sense organs in the retina of the eye that are sensitive to dim light.

Roost - a branch or other perch on which a bird settles for rest.

Scavenge - to feed on dead animals or other decaying matter.

Secretes - produces and releases.

Tentacles - the thin, flexible parts that extend from the body of an octopus, jellyfish, or other animal. Tentacles are used for grasping and moving.

Tuber - short, thick underground stem.

Ultrasonic - of sound waves having a pitch above the upper limit of human hearing.

Venom - the poison that some snakes, spiders, scorpions, insects, or other creatures can transfer through a bite or sting to a person or another animal.

Vibrations - rapid movements back and forth.

INDEX

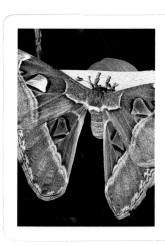

Sugar Glider

Scientific Name: *Petaurus breviceps*
Habitat: Forest or woodland
Range: New Guinea and eastern coastal Australia
Diet: Insects, larvae, small birds, and blossoms
Size: 5-12 inches (12.5-30 cm)

Jaguar

Scientific Name: *Panthera onca*
Habitat: Forest, marshes, scrublands, grasslands
Range: Mexico, northern Argentina
Diet: Rodents, peccaries, deer, tapirs, fish, cattle
Size: 3.75-6 feet (1.1-1.8 m) in length

Dragon-headed Katydid

Scientific Name: *Eumega lodon*
Habitat: Forests
Range: Malaysia
Diet: Plants
Size: About 3 inches (7.5 cm)

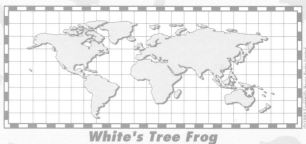

White's Tree Frog

Scientific Name: *Litoria caerulea*
Habitat: Woodlands, shrubby areas, trees close to water
Range: Northeastern Australia, islands of Torres Strait, and New Guinea
Diet: Large insects, other invertebrates
Size: 2-4 inches (5-10 cm)

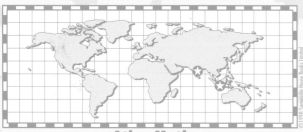

Atlas Moth

Scientific Name: *Attacus atlas*
Habitat: Tropical and subtropical forests
Range: India to Malaysia and Indonesia
Diet: Caterpillars feed on leaves of diverse shrubs and trees
Size: 6-12 inches (15-30 cm)

Wolf

Scientific Name: *Canis lupus*
Habitat: Tundra, steppe, open forest, mountain, swamp, grasslands
Range: Canada, US, Mexico, Europe, northern Asia
Diet: Deer, elk, moose, beaver, mountain sheep, mice, fish, carrion
Size: 3.5-5 feet (1-1.5 m)

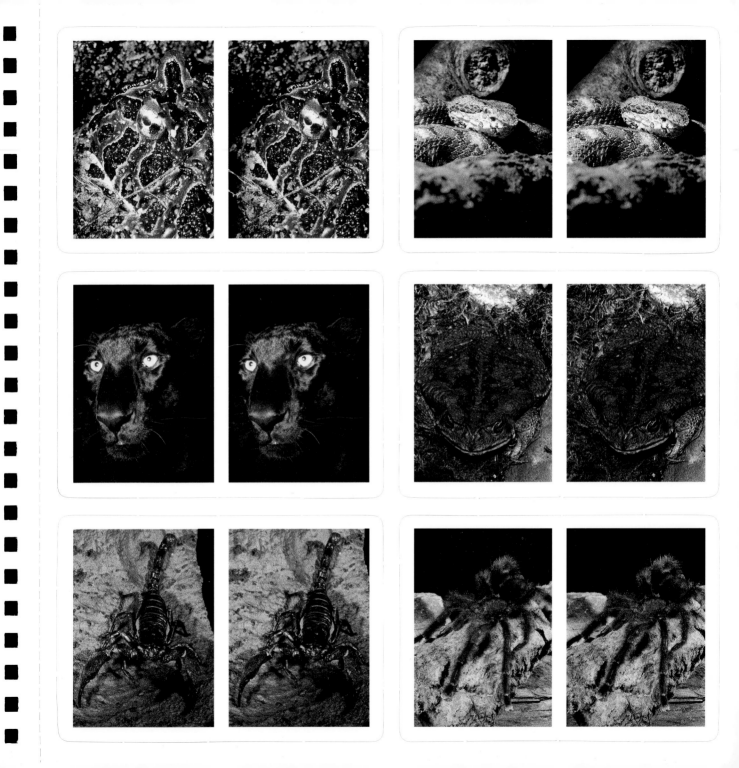

Eyelash Viper

Scientific Name: *Bothrops schlegeli*
Habitat: Tropical rain forests
Range: Southern Mexico to Ecuador and Venezuela
Diet: Lizards, frogs, small birds and mammals
Size: 16-24 inches (40-60 cm)

EYE-TO-EYE BOOKS

Bell's Horned Frog

Scientific Name: *Ceratophrys ornata*
Habitat: Tropical rain forests
Range: Argentina, Uruguay, Brazil
Diet: Insects, small amphibians, and mammals
Size: 3.5-6 inches (9-15 cm)

EYE-TO-EYE BOOKS

Marine Toad

Scientific Name: *Bufo marinus*
Habitat: From savannas to rain forests, cultivated lands, and suburbia
Range: Extreme southern Texas to Patagonia
Diet: Insects, arachnids, small rodents
Size: 4-9.5 inches (10-24 cm)

EYE-TO-EYE BOOKS

Leopard

Scientific Name: *Panthera pardus*
Habitat: Arid semidesert to dense rain forest
Range: Central and eastern Africa, but some in West Africa and Asia
Diet: Antelope, deer, pigs, domestic livestock, birds, rodents
Size: 3-6.25 feet (90-188 cm)

EYE-TO-EYE BOOKS

Pink-toed Tarantula

Scientific Name: *Avicularia avicularia*
Habitat: Jungles and pineapple and banana plantations
Range: Trinidad to Brazil
Diet: Katydids, small lizards, birds
Size: About 1.5 inches (4 cm)

EYE-TO-EYE BOOKS

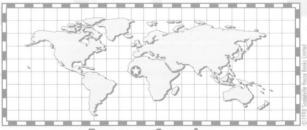

Emperor Scorpion

Scientific Name: *Pandinus gambiensis*
Habitat: Rain forests
Range: West Africa
Diet: Insects and carrion
Size: 4-6 inches (10-15 cm)

EYE-TO-EYE BOOKS

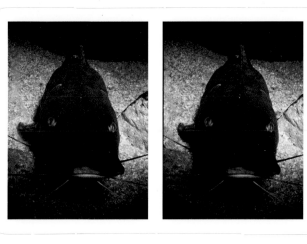
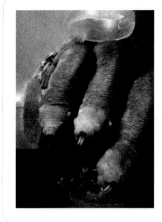

Naked Mole-rat

Scientific Name: *Heterocephalus glaber*
Habitat: Dry steppes and savannas
Range: Ethiopia, Somalia, Kenya
Diet: Mostly tubers and roots, sometimes insects
Size: 3.25-4.5 inches (8-11 cm)

EYE-TO-EYE BOOKS

Shovelnose Catfish

Scientific Name: *Pseudoplatystoma fasciatum*
Habitat: Large rivers and deeper sections of smaller rivers
Range: Northern and eastern South America
Diet: Fish, crustaceans
Size: About 3 feet (1 m)

EYE-TO-EYE BOOKS

Margay

Scientific Name: *Leopardus wiedii*
Habitat: Forests
Range: Northern Mexico, east of the Andes to northern Argentina
Diet: Rodents, birds, reptiles
Size: 18-31.5 inches (45-79 cm)

EYE-TO-EYE BOOKS

Raccoon

Scientific Name: *Procyon lotor*
Habitat: Forests, meadows, swamps, desert canyons
Range: Southern Canada to Panama
Diet: Crayfish, crabs, frogs, fish, nuts, seeds, acorns, berries, refuse
Size: 16-24 inches (40-60 cm)

EYE-TO-EYE BOOKS

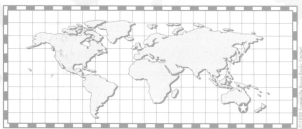

Southern Hairy-nosed Wombat

Scientific Name: *Lasiorhinus latifrons*
Habitat: Dry, open country, burrows
Range: Restricted areas of southern South Australia
Diet: Mainly grass, also roots and bark
Size: 30-40 inches (75-100 cm)

EYE-TO-EYE BOOKS

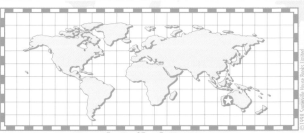

Brush-tailed Bettong

Scientific Name: *Bettongia penicillata*
Habitat: Brush woodlands, dry forest
Range: Southwestern western Australia
Diet: Mainly plants, tubers, roots, sometimes meat
Size: 11-16 inches (28-40 cm)

EYE-TO-EYE BOOKS

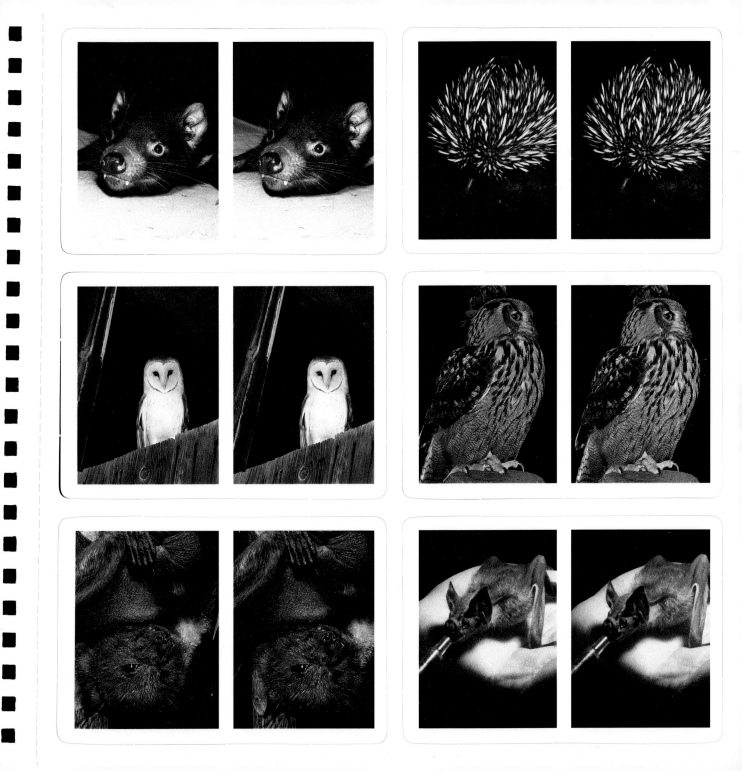

Short-beaked Echidna

Scientific Name: *Tachyglossus aculeatus*
Habitat: Forests, rocky areas, hilly tracts and sandy plains
Range: Australia, southern New Guinea, Tasmania
Diet: Ants and termites
Size: 14-20 inches (35-50 cm)

EYE-TO-EYE BOOKS

Tasmanian Devil

Scientific Name: *Sarcophilus harisii*
Habitat: Forests near the coast
Range: Tasmania
Diet: Carrion, insects, larvae, and vegetable matter
Size: 21-30 inches (53-75 cm)

EYE-TO-EYE BOOKS

Eagle Owl

Scientific Name: *Bubo bubo*
Habitat: Forests, rocky areas, woods
Range: Europe, Asia, and northern Africa
Diet: Mammals, birds, snakes, lizards, frogs, fish, large insects
Size: Length 25-29 inches (63-73 cm)

EYE-TO-EYE BOOKS

Barn Owl

Scientific Name: *Tyto alba*
Habitat: All forest habitats except boreal; all open, urban and residential habitats
Range: North and South America, northwestern Europe
Diet: Rodents, some small birds
Size: Length 14-20 inches (35-50 cm)

EYE-TO-EYE BOOKS

Vampire Bat

Scientific Name: *Desmodus rotundus*
Habitat: Open habitats
Range: Northern Mexico to central Chile, Argentina, and Uruguay
Diet: Blood of horses, burros, cattle, and occasionally humans
Size: 2.5-3.5 inches (6.5-9 cm)

EYE-TO-EYE BOOKS

Gray-headed Flying Fox

Scientific Name: *Pteropus poliocephalus*
Habitat: Forests and mangroves
Range: Eastern Australia
Diet: Nectar and juice from flowering and fruiting plants
Size: 9-11.5 inches (23-29 cm)

EYE-TO-EYE BOOKS